Trifon Angelov

The Parabens, Very Close To The Ideal Drug Preservatives

Trifon Angelov

The Parabens, Very Close To The Ideal Drug Preservatives

Determination of some physicochemical characteristics and antioxidant activity of parabens

LAP LAMBERT Academic Publishing

Impressum / Imprint

Bibliografische Information der Deutschen Nationalbibliothek: Die Deutsche Nationalbibliothek verzeichnet diese Publikation in der Deutschen Nationalbibliografie; detaillierte bibliografische Daten sind im Internet über http://dnb.d-nb.de abrufbar.
Alle in diesem Buch genannten Marken und Produktnamen unterliegen warenzeichen-, marken- oder patentrechtlichem Schutz bzw. sind Warenzeichen oder eingetragene Warenzeichen der jeweiligen Inhaber. Die Wiedergabe von Marken, Produktnamen, Gebrauchsnamen, Handelsnamen, Warenbezeichnungen u.s.w. in diesem Werk berechtigt auch ohne besondere Kennzeichnung nicht zu der Annahme, dass solche Namen im Sinne der Warenzeichen- und Markenschutzgesetzgebung als frei zu betrachten wären und daher von jedermann benutzt werden dürften.

Bibliographic information published by the Deutsche Nationalbibliothek: The Deutsche Nationalbibliothek lists this publication in the Deutsche Nationalbibliografie; detailed bibliographic data are available in the Internet at http://dnb.d-nb.de.
Any brand names and product names mentioned in this book are subject to trademark, brand or patent protection and are trademarks or registered trademarks of their respective holders. The use of brand names, product names, common names, trade names, product descriptions etc. even without a particular marking in this works is in no way to be construed to mean that such names may be regarded as unrestricted in respect of trademark and brand protection legislation and could thus be used by anyone.

Coverbild / Cover image: www.ingimage.com

Verlag / Publisher:
LAP LAMBERT Academic Publishing
ist ein Imprint der / is a trademark of
OmniScriptum GmbH & Co. KG
Heinrich-Böcking-Str. 6-8, 66121 Saarbrücken, Deutschland / Germany
Email: info@lap-publishing.com

Herstellung: siehe letzte Seite /
Printed at: see last page
ISBN: 978-3-659-60671-7

Zugl. / Approved by: Sofia, Medical University, Diss., 2009

Table of contents:

The Parabens - Very Close to the Ideal Drug Preservatives

Chapter I

Introduction

Among the most abundant definitions of drug preservatives is that they are compounds having bactericidal, bacteriostatic or fungicidal action [1].

The function of the preservative in a pharmaceutical preparation is to prevent microbial contamination.

Depending on their action, preservatives are divided into the following three primary classes [2]:

1. Antimicrobials

2. Antioxidants

3. Chelating agents

The ideal preservative needs to posses all three of the preservative properties mentioned above simultaneously. Under the certain conditions, the preservative can act as an antimicrobial composition, an antioxidant and a chelating agent.

In the beginning of XX century the term preservative was limited to antimicrobial agent. Sodium benzoate was the first chemical preservative permitted in food for human consumption in the U.S. in 1908, and has continued to be used in a large number of foods [3]. It belongs to the group of preservatives with an antimicrobial activity. Today ''preservative'' interprets any added substance whose role is to keep the product and its properties unchanged in the product's shelf life time [2][4][5].

The antioxidants also belong to the group of preservatives. Antioxidant supplements are added to prevent or delay the oxidation-reduction reactions and their unacceptable effects [6].

The role of the chelating agents is to preserve the homogeneity of the liquid drug products. This is particularly important for the uniformity of their composition. Especially essential for the emulsions and the suspension solutions is the role of the chelating (stabilizing) agents.

The ideal preservative must possess a number of qualities, such as antimicrobial activity against a large number of microorganisms, chemical and physicochemical stability, compatibility with the other ingredients of the product and its packaging. It also needs to cause no irittation and to be generally harmless to the health. Sometimes, the antimicrobial agents are used in combination in order to achieve an enhancement to their bactericidal and bacteriostatic activity against certain types of microorganisms, even at lower concentrations. With the simultaneous use of certain preservatives and the addition of several compounds in order to promote their anti-microbial action such a synergistic effect can reduce the level of their concentrations and thus reduce their toxicity [7].

In the selection of the preservative must be taken into account:

(a) toxicity and irritative action of the preservative

(b) pH range with maximum antimicrobial activity

(c) compatibility with the other ingradients of the product

(d) sustainability during the conditions of manufacturing

(e) synergism or antagonism in antimicrobial activity

Antioxidant activity

The addition of antioxidants is a practice in the production of pharmaceutical products. Among the most common are vitamins E and C, butylated hydroxyanisole (BHA), butylated hydroxytoluene (BHT), gallates and others. There is not extensive amount of data in the scientific literature about the antioxidant activity of parabens. The evaluation of parabens such as traps for the hydroxyl radicals OH * has been made in some tests with the Fenton reagent [8] [9] as well as by means of GC/MS and HPLC with an electrochemical detector in the in vivo study [10].

Casini at al. (1981) observed a delay in the process of photodegradation of riboflavin medicinal products preserved with methyl paraben. This is indicative of the possibility of parabens to capture free radicals initiated by light and is evidence of the antioxidant properties of esters of p-hydroxybenzoic acid [11].

Parabens

Overview

The parabens are among the most useful drug preservatives. For the first time they have been used at the middle of twenty years of the last century as drug preservatives in the production of suppositories, syrups, injectable solutions, eye drops and contraceptive products [12].

Among the parabens metyl paraben (nipagin) and propyl paraben (nipasol) are the most popular and useful drug preservatives.

In nature, methyl paraben naturally occurs in wild blackberry, about 0.15 ppm, in papaya juice, red and white wine [12].

The usable concentrations for the methyl paraben are: 0.065-0.250 % for injectable solutions, 0.015-0.050 % for ophthalmological products and 0.015-0.20 % for syrups and suspensions [12].

Propylparaben is used alone or in combination with other parabens or antimicrobial agents. It is not recommended for propylparaben to be used as the only preservative

4

in eye drops, as its effective concentration is going to irritate the eyes. To avoid this irritating action propylparaben is usually used in eye drops simultaneously with methylparaben. The synergism [12] of the simultaneus use of both parabens allows these preservatives to be more effective against a wide range of micro-organisms even at lower concentrations. The ratio between the concentrations of both parabens is in favor of the methylparaben. Usually methylparaben is used in concentrations from two to four times higher than the concentrations of propylparaben and depends on the product`s media. For drugs the maximal concentration of parabens usually does not exceed 1% [12].

The presence of some substances can affect of the biological action of parabens. Some authors demonstrated that syrups could be preserved with parabens and that the addition of small amount of propylene glycol (usually 2% to 5%) potentiated the antimicrobial activity of the parabens. There is publication which reports that parabens can be employed as efficient preservatives in the presence of nonionic surfactants, but the proper concentration must be employed [1]. Hydrophobic esters, such as propyl and butyl are usually effective at lower concentrations; but in the presence of surfactants of the polyoxyethylenlene type, this relationship may be reversed [1].

In the preparation of suspension it is often necessary to add an inert suspending agent to retard sedimentation. To enhance the physical stability of the suspension, suspending and emulsifying agents may be added. Such substances are gum tragacanth or gum acacia, as well as the cellulose derivatives [1]. These substances are subjected to attack by microbes. This would certainly indicate that the presence of antimicrobial agents becomes necessity for these types of products. Having determined the necessity of an antimicrobial agent in a formulation it is possible to be faced the problem of the formation of complexes of the cellulose with number of preservatives. Some authors reports that methyl cellulose forms complexes with parabens, and the degree of interaction decrease in order: methylparaben, propylparaben and butylparaben [1].

5

Mechanism of antimicrobial action

Unlike most week organic acids, unpopular compounds such as parabens show very slight pH dependency on the environment, and the mechanism of their antimicrobial action is also different from the antimicrobial action of the weak organic acids.

Table 1. The ratio in percent of the undissociated forms of weak organic acids at different pH ranges.

pH dependency of the dissociated constant of some of weak organic acids					
pH range ＼ Organic acid	3	4	5	6	7
Acetic acid	98.5	84.5	34.9	5.1	0.54
Benzoic acid	93.5	59.3	12.8	1.44	0.144
Citric acid	53.0	18.9	0.41	0.006	<0.001
Parabens	>99.99	99.99	99.96	99.66	96.72
Propionic acid	98.5	87.6	41.7	6.67	0.71
Sorbic acid	97.4	82.0	30.0	4.1	0.48

Using hydrophobic parabens is one of easiest ways for disrupting the normal function of mitochondrias. This action is accompanied by depolarization of the mitochondrial membrane [13] and lowering of the oxidative phosphorilation. The disorder in the mitochondrial function leads to cell death. It is considered that the antimicrobial action of parabens also depends on their lipophilicity, which allows them to penetrate and to be retained in the cellular membrane. In this way parabens damage the double lipid layer and affect the transport processes of the cell [14]. Furthermore this can lead to distortion of the mechanical integrity of the cellural membrane [15].

Depending on their impact on the microflora, preservatives are divided into two groups – agents with bactericidal and bacteriostatic action [16][7]. Preservatives with bactericidal action destroy the microorganisms present in the product, and preservatives with a bacteriostatic - prevent the growth and propagation of the already present, and maintain the quantity of the microorganisms within the human safety standards. [16][7].

The esters of the p-hydroxybenzoic acid are more fungistatic than fungicidal. Parabens are effective to the germination of spores and development rather than the vegetative growth of bacteria and molds. The fungicidal action of methyl paraben is more pronounced than the bactericidal [12]. The activity of the methyl paraben against gram positive microorganisms is greater than that relative to the negative. The increasing in the length of the hydrocarbon chain in the ester group causes lowering of the solubility of the parabens in water [12].

Figure 1. Biochemical degradation of parabens

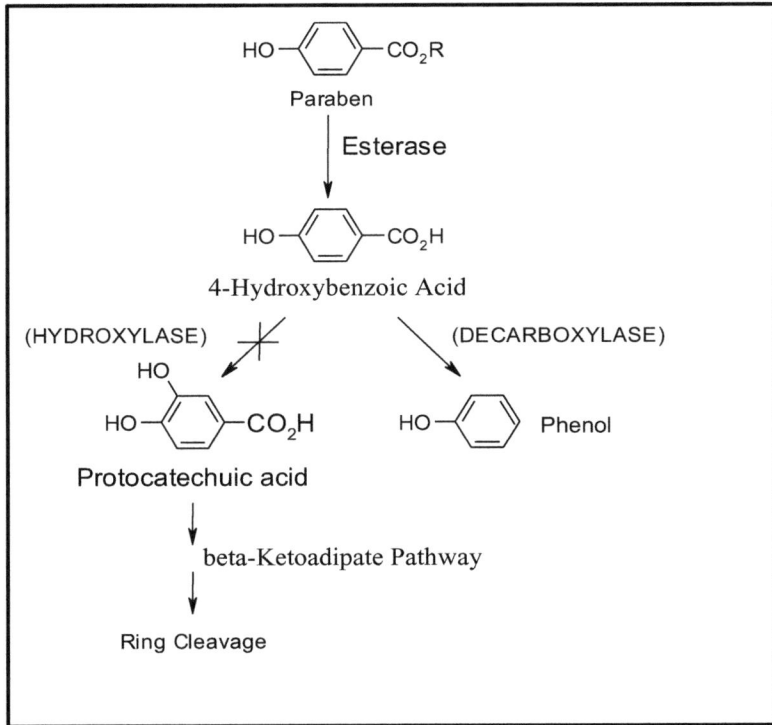

The bacterial strains Enterobacter EM are resistant to methyl and propyl parabens. They can survive in an aqueous medium containing nipagin and nipazol in concentrations of 1700 and 180 mg.l^{-1}, or 11.2 and 1.0 mol.l^{-1}, as well as at very high concentrations of the methyl paraben 3.000 mg.l^{-1} or 19.7 mol.l^{-1} [13]. These micro-organisms can hydrolyse about 500 mg of parabens in about 2 hours. (Figure 1).

The molecules of propylparaben have better lipophilicity than the molecules of the methyl paraben, penetrate the cellular membrane of the microbial cell more easily and depolarize the mitochondrial membrane more quickly [13][17].

Metabolism and toxicology

Orally administered methyl paraben [12] is absorbed easily and completely in the gastrointestinal tract and is hydrolyzed to the p-hydroxybenzoic acid under the action of the enzyme esterase. Metabolites of methyl paraben in urine appear within half an hour after dosing [12]. Identified metabolites are p-hydroxybenzoic acid, 4-hydroxy-hippuric acid, ester and ether glucuronides. Methyl paraben is not detected in the urine. Tasukamoto and Terada (1960, 1962, 1963) investigated metabolites of methyl and propyl paraben in rabbits and found that 39% of the orally administrated nipagine was excreted as p-hydroxybenzoic acid, and the rest - 15%, 4-hydroxy-hippuric acid (Figure 2), 21% as esters and ethers of glucuronic acid and 10% sulfuric acid conjugates of p-hydroxybenzoic acid [24].

In single doses of 0.4-0.8 g.kg-1 in 24-hour 86% of the orally administrated amount of methyl paraben is excreted by urine [12].

Nipagine and nipazol increase the activity of the enzyme dihydrofolate reduktaze. It is considered that this is due to the fact that methyl and propyl paraben induce conformational changes in the structure of the enzyme, which increases its affinity to dihydrofolates [17]. The connections of parabens with bovine serum albumin are reported in literature [12]. The extent of the binding increases with the increasing of the length of the hydrocarbon chain in the ester group [12][17].

In humans, about 30% of the oral nipagin is adsorbed in the mouth. The absorption of parabens in the oral cavity increases with the increasing of the lipophilicity of the parabens, i.e. methyl paraben has got the smallest absorption.

Figure 2. Metabolic pathway of parabens to 4-hydroxy-hippuric acid [12][17].

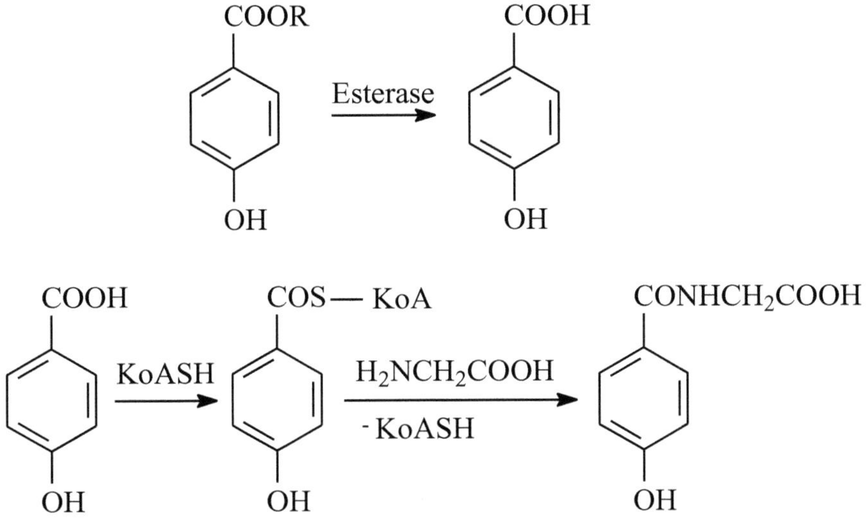

$$COOR \xrightarrow{\text{Esterase}} COOH$$

(benzene ring with OH para to COOR) → (benzene ring with OH para to COOH)

$$COOH \xrightarrow{\text{KoASH}} COS-KoA \xrightarrow[\text{- KoASH}]{H_2NCH_2COOH} CONHCH_2COOH$$

(benzene ring with OH para to COOH) → (benzene ring with OH para to COS—KoA) → (benzene ring with OH para to CONHCH$_2$COOH)

Propyl paraben and its metabolites do not accumulate in the body like methyl paraben. Propyl paraben is rapidly hydrolyzed by esterases and is metabolised in the same way as the methyl paraben in the liver and kidneys [18]. The main metabolites of nipazol are p-hydroxybenzoic acid, p-hydroxyhippuric acid, ester and ether glucuronide and sulfate conjugates [17].

The preservatives belong to the group of the hazardous substances. It is essential to know under what conditions most of the toxic effects of the preservatives occur. In susceptible individuals this may lead to more serious consequences such as sensibilization of the patient and even anaphylactic shock [12][17][19][20][21]. Most of the preservatives are contained in natural products as a variety of natural compounds [22][23]. In many plant products they are found in the form of salts and esters [24], and produce no irritating effect to most people.

Clinical studies with humans have found that parabens are generally irritating. The results of the study of Veinen at al. (1996) found that the combination of methyl and propyl paraben may cause dermatitis and eczema [25]. Bajaj and Chattergee (1985) found that some people have strong hypersensitivity to methyl paraben [26]. A mixture of 0.02% nipasol and 0.04% nipagin induces mild inflammation of the eye and the aqueous solution of methyl paraben at a concentration of 0.1-0.3% causes slight watering of the eyes and redness, but all symptoms disappear within 1-2 minutes. Widely discussed in literature is the so-called "paraben paradox". Fisher (1996) found that when parabens penetrate the skin, they may cause dermatitis, but when administered orally or injected, this effect is not observed. The explanation of this paradox can be given by the Langerhans cells and their role in explaining the acute dermatitis [20]. From the toxicological studies it can be concluded that methyl paraben is the least toxic paraben. It has very low acute and chronic toxicity. The acceptable daily intake for nipagin defined by FAO/WHO is 55 mg.kg^{-1} body weight per day.

Propyl paraben has got a weak chronic toxicity, but is more toxic than the methyl paraben. The acceptable daily intake for nipasol defined by FAO/WHO is 10 mg.kg^{-1} body weight per day, but in view of its low toxicity it has been upgraded to that of the nipagine - 55 mg.kg^{-1} body weight per day [17][27].

One of the studies with experimental animals has shown another of toxicological effects of the propyl paraben, which is not characteristic for the methyl paraben. When the consumption of propyl parahydroxybenzoate is increased it affects the amount of testosterone and the concentration of sperm in the semen. This has been found by Oishi (2002) [27].

Parabens are characterized by low toxicity, which is expressed mainly by their irritating action which disappears quickly. In more sensitive patients, however, it can also be seen in acute reactions, especially at high concentrations of the preservatives.

Methods of analysis

For the analysis of parabens the most applicable are the spectrophotometric and the chromatographic methods [5][28][29][30][31][32][33][34][35][36][37][38]. The method of Collado at al. (2000) for determination of several preservatives includes also the determination of nipagin in eye drops [39]. Besides the methods mentioned here the methods of micellar electrokinetic chromatography [40][41] and capillary electrophoresis [41][42][43] are also increasingly encountered in the scientific literature. The choice of the appropriate method for a particular type of analysis is a function of the purpose of the analysis and the available equipment.

Nipasol is often used as an internal standard in many liquid chromatographic determinations [44]. As an example can be mentioned the determination of hydrocortisone, methotrexate, meperidine, melphalan, cytarabine and clindamycin and others [44].

Knowledge of the physicochemical characteristics of preservatives can significantly help to determine the optimal concentrations of the application. Such characteristic is their lipophilicity. Data on the value of this parameter for the various conditions is sufficiently precise to limit the effective concentration of preservatives with the aim to reach a new, lower concentration level of the preservative in the drug.

Lipophilicity. Index of lipophilicity

Chromatographic methods may also be used for determining certain physicochemical properties of the substances [45][46][47]. Such characteristics are the lipophilicity parameter and pKa value of the analyts [48]. In this case, the chromatographic methods can be regarded as absolute, i.e. the obtained results are not compared with the results obtained from a reference standard. This type of applications of chromatographic methods, however, in most cases requires the use of a complex mathematical apparatus and the development of a methodology for the evaluation and interpretation of the obtained results. Due to the dynamic mode the chromatographic methods do not always supply sufficiently accurate results, but in most cases are faster than the classical methods used for the same purpose [45]. In some cases, however, under the suitable chromatographic conditions the dynamic mode of the chromatographic methods has an advantage because it allows us to model the naturally occurring processes and to extract valuable information about the mechanism of this processes. The result of such an examination is very close to the actual one in comparison with the result obtained for the same parameter in another method. An example of such a test is the determination of the parameter of the lipophilicity of the substances [48].

General concepts. Classical method for determining of the lipophilicity parameter

The lipophilicity parameter does not have an exact definition. In general, lipophilicity is often expressed (not very correct) as hydrophobicity. The hydrophobicity of a compound is defined as its relative tendency to be readily soluble in non-polar solvents and only sparingly soluble in water [48]. Since the process of dissolution is dependent on the parameters of the very molecules, from various studies made for the determination of the lipophilicity it can be concluded that the concept of lipophilicity can be expressed as a function of the bulk (excluded volume) of the molecules and their polarity [49]. When expressing the lipophilicity as a function of bulk properties of the molecules, the non-specific and mainly indirect Van der Waals forces and the hydrophobic interactions need to be taken into account. They are also considered to be the main reason for the increase in the entropy of the system. Expressing the lipophilicity as a function in polarity terms, the electrostatic interactions as well as the interactions caused by the nature of the molecules of the solvent-solute (dipole-dipole, ion-dipole), and the formation of hydrogen bonds need to be considered. It is believed that these forces are related to the change in enthalpy of the system [49].

The lipophilicity of the preservatives is an important characteristic in the determination of their antimicrobial activity. It is one of the main parameters for the first stage of the antimicrobial action, namely the penetration and introduction of the preservatives in the bacterial cell and the cellular wall. Once penetrating the cell through the cellural membrane, the molecules of the antimicrobial agent can easily pass subsequently through the membrane of the individual organelles (e.g. mitochondria) [50]. Because of the complicated structure of the cell membrane, when passing through the different parts, the molecules have to make a lot of passes between the water / organic interfaces, the cell cytoplasm and eventually (but not in all cases) in the membrane of the different cell organelles. The process is much more complex than a simple partitioning in the interfacial boundaries, as it is accompanied by many acts of adsorption and desorption on solid surfaces of the parts of the cell membrane. The acts of adsorption and desorption are parallel to the passing of the molecules in the interfacial boundaries and the relative polarity of the molecules plays a major role in these elementary acts. The process of partitioning in the interfacial boundaries depends on the lipophilicity of the molecules, and the process of adsorption and desorption depends on the polarity of the molecules. The molecule for which the sum of free energy changes is minimal for the many boundary crossing of the two types of processes (adsorption and desorption) will be with ideal lipo-hydrophilic character. This molecule will reach most easily the subject of its biological activity [50].

The classical method for the evaluation of the lipophilicity parameter of a compounds is by determination of the coefficient of distribution in the n-octanol/water system [48] [51]. The selection of this particular system has been chosen by Hnsh and Fudjita in order to emphasize its density, viscosity, dielectric constant, water solubility and high process structure saturated with water molecules octanoic phase (each water molecule coordinated around four molecules of n-octanol) [49]. The method is based on a liquid-liquid extraction (shake flask techniques) in the system n-octanol/water. By determining the concentrations of the investigated compounds in both phases of the system, the partition coefficient in water and n-octanol is determined.

$P = [organic]/[aqueous]$

where [organic] is the concentration of the compound investigated in n-octanol lyer [52].

where [aqueous] is the concentration of the compound investigated in water [52].

It is accepted that lipophilicity should be expressed with the decimal logarithm of the partition coefficient in the octanol/water system [51].

LogP = log10 (Partition Coefficient)

The partition coefficient (P) describes the ability of neutral (uncharged) molecules to be dissolved and partitioned between the different phases in two-phase systems organic part (fats, oils, organic solvents) and water, or how much is soluble the test substance in one fraction compared to the other. Substances which are readily dissolved in the aqueous part of the system are called hydrophilic and those which preferentially dissolve in the organic portion are called hydrophobic or lipophilic, when the organic layer is a lipid. It is very important to know the partition coefficient of the substance as it provides important information about the physical nature of the substance and allows us to predict its behavior under different conditions. The logarithm of a partition coefficient (logP) gives us an indication of whether the substance is absorbed by the plant, animal, human or other tissues or is easily excreted in water [52] [53].

The negative value of logP indicates that the test substance has a greater affinity for the aqueous phase and is therefore hydrophilic. When logP = 0 the distribution between the aqueous and lipid phases is equal. The positive logP indicates that the substance is in a higher concentration in the lipid phase, and therefore the compound is more lipophilic [48][49][51][52]. The lipophilicity is very important for the absorption of the componds, their distribution in the body and partitioning in the cell membranes and the biological barriers.

The distribution coefficient of a substance can be determined experimentally by a variety of methods. As mentioned above, the most common method among them is the liquid-liquid extraction, but in the last years the method of high performance liquid chromatography is more widely used. The method of liquid-liquid extraction is suitable for a large number of compounds. It is very labor intensive and requires long experimental time, but is considered to be more accurate. Among the achievements of the method of the high-performance liquid chromatography (HPLC) are its rapidity, and the possibility of being used for determination of the lipophilicity of the compounds with known structure and without high purity.

If the investigated compound contains one or more ionizable groups, it could exist as a mixture of one or more ionic forms. The composition of this mixture strongly depends on the pH of the medium. Since log P is defined only for neutral components, [52] in the case of ionization it is recommended that the effective distribution coefficient D should be used as the distribution coefficient for the mixture of the ionic and neutral form. For the system octanol/water, D and its corresponding logarithmic form log D [52] give us the correct description of the system by the equilibrium of the different forms of the compound. Log D is responsible for the correct (true) behavior of ionizable compounds in the defined pH ranges, considering all ions present. This parameter is very useful, especially in assessing pH-dependent pharmacokinetic properties as bioavailability, metabolism, and toxicity.

The relation between log$P$ and Log D is given by the equation (1) [49]

Equation (1)

log$P$ = Log D + correlation factor

for monobasic acids and bases the correlations factors are respectively:

$$\log\left(1+10^{pH-pKa}\right) \text{ and } \log\left(1+10^{pKa-pH}\right).$$

Determination of the lipophilicity parameter using reversed phase high-performance liquid chromatography (RP-HPLC)

With the development of the method of the reverse phase the high performance liquid chromatography (RP-HPLC) started to be widely used in order to determine the lipophilicity parameter, as an analogue of the partition coefficient in the system n-octanol/water. In RP-HPLC the retention factor is used as a measure of the lipophilicity. Earlier it was known as a capacitive factor, and later as the retention factor k' of the column of the test compound [54].

The distribution of the test compound between the stationary and mobile phase in a liquid chromatographic separation is a result of the existing forces between the molecules of the analyte and the molecules in each phase [56].

Equation 2. Retention factor

$$k' = \frac{n_s}{n_m} = \frac{c_s V_s}{c_m V_m} = \frac{(R_t - R_0)}{R_0} \tag{2}$$

n_s- mols of analyte in the stationary phase, n_m- mols of analyte in the mobile phase, c_s - concentration of analyte in the stationary phase, c_m -concentration of analyte in the mobile phase, V_s - volume of the stationary phase, V_m – volume of the mobile phase, R_t - the retention time of analyte and R_0 is the retention time of unretained compound [54][55].

In the reverse phase of the high-performance liquid chromatography, the retention of the investigated compounds is due to the hydrophobic forces between the ligands of the stationary phase and the unpolar fragments of the molecules of the compounds [56]. The retention data can be used in order to calculate the lipophilicity parameter - log P [57] and can be compared with the data obtained by the classic method of liquid-liquid extraction. There are three main approximations of the correlation between the chromatographic data and the data obtained by the classic method of liquid-liquid extraction [57]:

In the first approximation for obtaining the data for the log k' is used a specific chromatographic column with a defined mobile phase.

In the second approximation is used log k'w. The physical meaning of log k'w is the capacitive factor k', determined by using the mobile phase containing only water without an organic modifier. The correlation between the log k' and the log k'w is given by equation 3 [55]:

$$\log k' = \log k_w - S\varphi \qquad \textbf{Equation (3)}$$

where k_w corresponds to the capacitive factor in isocratic elution k' with eluent pure water and is usually an extrapolation parameter, φ is the percentage of the organic modifier in the mobile phase, and S is related to the elution strength of the pure organic modifier.

The third approximation suggested back extrapolation method for the determination of the log k', recommending optimal content of the organic solvent in the mobile phase so that the system n-octanol / water can be modeled most accurately [57].

As the value log k'w in the second approximation can be directly determined only to a small number of components, it is required that a correlation between the content of the organic solvent in the mobile phase and the log k' should be found, so that log k'w can be predicted [57]. However, the results show that the linear correlation is not valid for a wide range of concentrations of organic modifiers, and the obtained data for log k'w is not identical when a variety of organic modifiers (methyl alcohol and acetonitrile) is used. Some authors suggest the use of a quadratic dependence in predicting log k'w, while others use the solvofobic theory of Horvath et al. for predicting the log k'w [57].

Considering the behavior of the stationary phase of the chromatography column as a liquid, the capacitive factor for an analyte is related to the characteristics of the chromatographic system through the connection:

$$K = D * \frac{Vs}{Vm} \quad \textbf{Equation (4)}$$

D is the partition coefficient of the test substance in the volumes of the stationary and the mobile phases, respectively **Vs** and **Vm** [49]

It is believed that due to the balanced lipo-hydrophilic properties of normal octanol, it describes the behavior of molecules in the biosystems most accurately. In fact the normal octanol is an isotropic liquid, which is in contrast to the high anisotropy of the bio-membranes. The main components of most membranes are the phospholipid and cholesterol molecules which form a layer in which protein molecules and lipids are incorporated. In general, the penetration of biologically active substances through the cell membrane is determined by three structurally different regions. The outer region is composed of the heads of the phospholipid groups and is polar. The next region is with middle polarity, and presents a water organic interface containing cholesterol, a glycerol skeleton and the first few methylene groups of the phospholipids. The last region consists of the tails of phospholipids [48]. The last region is not highly polar, loose and very flexible. From the description of the membrane it can be concluded that the distribution of components between the regions is not similar to the distribution between the two liquids, but rather the size, the shape of its molecules, and the type and location of functional groups in the different structures of the membrane will be of importance for the distribution coefficient of the substances in the cell wall [48]. The structures of the reversed phase chromatographic surface and the bacterial membrane are shown on the Figure 3.

Figure 3. Outer membrane of the representative of the family *Enterobacteriaceae* and the area of reversed phase silica column

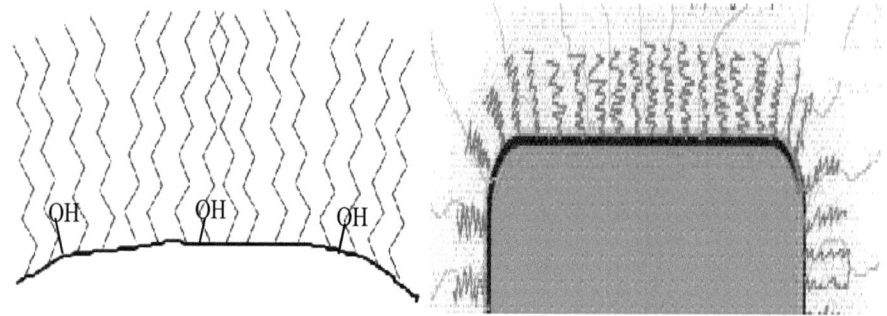

Considering the surface of a RP-HPLC column it is evident that there is similarity between the interface mobile phase/stationary phase and the interface of water/bio-membrane. The behavior of the stationary phase, which is a covalently bound hydrocarbon residue, is not so liquid and resembles the stacked array of hydrocarbon chains in the membrane [48]. The unmodified silanol groups, some of which are charged at neutral pH, absorbed on the surface and connected with the hydrogen bonding to the silanol groups organic modifier and the water molecules dissolved in this organic modifier, are very similar to the polar region of the outer membrane. Another similarity is that both the systems are in the dynamic mode, wherein equilibrium is rarely achieved [48]. The purpose of this comparisons is to focus on the specific interactions between the solute and the lipophilic phase, which usually exceed the hydrophobic effect in a RP-HPLC analysis. The retention parameters in the reversed phase liquid chromatographic system can carry additional information on the physicochemical properties of the tested components which cannot be obtained in the study of the distribution of the substance in the system of the two fluids. In most systems, the information may be further used for quantitative description of the hydrophobic nature of the interaction of bioactive components.

Other advantages of RPLC when determining the lipophilicity parameter in comparison to the method of liquid-liquid extraction are:

* Improved accuracy and reproducibility
* Determination of a wide range of values for the index of lipophilicity
* Reduction of the interference from impurities in the analytes
* The rate of determination as well as little solvent consumption [49].

Dependency on the lipophilicity of the analyte of pH of the medium

The lipophilicity of the components is expressed by the ratio of their distribution or logarithm of the distribution coefficient logP between a non-aqueous and an aqueous phase in the system n-octanol/water. As mentioned above, the system normal octanol/water was first proposed by Fujita et al. and Hansch et al. [58] and has gained the highest popularity. As the most of the substances in the pharmaceutical manufacturing are weak acids or bases their distribution in the human body depends on their lipophilicity and dissociation. Unfortunately the data logP and pKa are known for a number of stable substances [59]. This is the reason for the development of a fast and convenient procedures for determining logP and pKa.

The isocratic reverse phase liquid chromatographic determination of the pKa requires the conduct of several experiments at different pH values of the mobile phase, and the determination of the acidity of the mobile phase is done under the certain conditions (T°C). The retention of the ionic analytes in HPLC strongly depends on the pH of the mobile phase. The retention factor k (formerly known as capacitive factor) of the non-dissociated form of substances with an acidic or basic character, can be 10 to 20 times greater than the retention factor of the corresponding dissociated form at a constant ratio of water or a buffer/organic solvent in the mobile phase composition [55]. The retention factor k for ionic analytes vary within wide limits with the change of pH and a constant ratio of the organic component in the mobile phase. The relation between the log k and the percentage of the organic component of the mobile phase is linear [59]. The variation in pH is actually a convenient method for the rational modification of separation at ionizable analytes [60][61].

The RP-HPLC methods for determining the lipophilicity described in the specific literature report only data for the pH of the aqueous portion of the mobile phase but not in the mobile phase integrity. Some authors suggest that the measurement of the pH should be done after the mixing of the aqueous buffer solution with the organic fraction of the mobile phase, as this gives the correct pH value of the mobile phase as a whole [59]. The recommendations of the IUPAC are that pH measurement should be done after the mixing of aqueous buffer with the organic modifier, and that the electrodes for measuring the pH should be calibrated prior to that calibrated with an aqueous buffer or a buffer prepared in the same solvent used in the mobile phase [61].

Lewis et al. has fixed the underlying causes of errors in predicting the retention of the investigated compounds as a function of the pH:

1.   Interaction of the analyte with the free silanol groups or metal ions on the surface of the stationary phase.

2.   Influence of the effect of the ionic strength on the Ka value of the analyte.

3.   Influence of the effect of the solvation of the ionic strength on the retention.

4.   Ion Pairs interactions of the samples ions with the buffers ions and the ionized components of the buffer.

The retention of the neutral compounds also depends on the pH of the mobile phase, but this dependence is much weaker than the dependence of the ionizable components [59].

Knox et al. investigated the effect of the ionic strength of the eluent on the retention factor of the anionic analytes with carbon modified silica stationary phases [59]. They reported negative values of the retention factor k' of the ionized organic acids. The reason is that the ionized molecules of the organic acids are not retained in the pores of the sorbent. These pores are accessible for the molecules of the mobile phase and the molecules of the analyts are eluted from the column before the molecules of the mobile phase. The process of the exclusion of the molecules of the acids becomes most intense when the ionic strength of the eluent is in the limits of 0.1-1 M [59].

Knox et al. also demonstrate that the elution of the analytes anions from the modified reverse phase sorbents depends on a complex way of the composition of the eluent [59]. They describe a U-shaped curve dependence on the retention factor k' of a mobile phase composed of ethanol and water at different ratios for three simple organic acids. The fastest exclusion was observed in 50% of ethanol in the mobile phase. At a water content of more than 80% the benzoic and the salicylic acids are more strongly retained by the column while sulphanilic acid is still eluted with the dead volume of the column [59].

Very important for the biologically active molecules is the presence of ionic functional groups (usually carboxyl or amino) in their structure, so that the parameter of acidity pKa is of great importance for prediction of the physical, physicochemical and biological properties of the structurally related substances. The specificity in the pharmacokinetics of xenobiotic substances (absorption, metabolism, excretion, toxicity) depends on the acidity [47]. This is the reason why the modern concepts in the development of new drug products require a simple and easy way to determine the pKa of these substances. The pH metric titration and the spectrophotometric analysis are routine in the process of determining the parameters of the acidity. However, there are several limitations on the usage of these methods. Among them are the poor solubility at pH-metric titration [63], the lack of chromophores in the spectrophotometric determination, as well as the differences in the spectra of ionized and non-ionized forms. An inert water-soluble salt (0.15 M KCl or NaCl) is added to the solution in order to improve the measurement precision, and to mimic the physiological state [64]. As many of the components of interest to the pharmaceutical manufacturing are insoluble in water and are with insufficient purity, some of the methods mentioned above are not applicable [47]. This is the reason why in the last twenty years liquid chromatography is often used for the determination of dissociation constants of ionic substances [47].

The advantages of the liquid chromatography are that this method uses very small quantities of test substance and the possibility that the investigated substances do not have very high purity. Among the liquid chromatographic applications are used the methods of the reversed phase chromatography, ion chromatography, and recently the method of capillary electrophoresis. The use of the reverse phase chromatography (RP-HPLC) in order to determine the pKa [47] is based on the solvofobic theory of Horvath et al. This theory assumes that the chromatographic behavior of the substances with acidic and basic character and tsviterionnii components is determined by two main effects, the hydrophobic effect (the stationary phase) and the reverse ionization of the analyte in the mobile phase [46]. The electroneutral forms of analyte have greater affinity to the non-polar stationary phase and consequently have greater retention in comparison with the ion forms of the same analyte, which are more soluble in the mobile phase. That is the reason why the dissociation of ionogenic analytes which depends on their pKa value and the pH of the mobile phase have a strong influence on the chromatographic behavior of the analytes [46].

Horvath et al. Have started to investigate theoretically and experimentally the influence of the effect of ionization of weak acids, bases and ampholytes in retention on organically modified silica in 1977. The retention, according to this concept, is due to the reversible association of the molecules in the soluted analyte S over the hydrocarbon ligands L (mostly C_{18}) of the stationary phase [46].

$$S + L \rightleftharpoons SL \qquad \text{Equation (5)}$$

The equilibrium constant of this association K_{assoc} is expressed by the equation 6:

$$K_{assoc} = \frac{[SL]}{[S][L]} \qquad \text{Equation (6)}$$

When expressing the equilibrium constant of the association process K_{assoc} for the neutral and ionized molecules is meant only the solvating effect, without taking account of the ionic interactions and the hydrogen bonds between the analyte and the hydrocarbon ligands [46].

In the situations where the analyte is a weak monobasic acid HA, the equilibrium constant of this process is the dissociation constant of the acid under the conditions specified by the mobile phase,

$$K_{a_m} = \frac{[H^+]_m [A^-]_m}{[HA]_m}$$ **Equation (7)**

where $[H^+]_m$, $[A^-]_m$ and $[HA]_m$ are respectively the concentration of the protons and the concentrations of the dissociated and undissociated acidic molecules in the mobile face.

In the process of chromatographic determination the retention of the analyte is considered a reversible association between the molecules of the dissociated or the undissociated acid and the hydrocarbon ligands L of the stationary phase. Expressing the capacity factor by means of these equilibria and equilibrium constants, respectively to the dissociated and the undissociated acid, yields the equation 8:

$$k = \frac{k_0 + k_{-1} \dfrac{K_{a_m}}{[H^+]_m}}{1 + \dfrac{K_{a_m}}{[H^+]_m}}$$ **Equation (8)**

The relationship between the capacity factor, the concentration of hydrogen ions in the eluent, the dissociation constant and the two limits capacitive factor of undissociated and dissociated acid is clearly demonstrated (equation 8). The detailed description of the methodology of calculation can be found in [46].

Like the weak acids and the weak bases dissociation in the mobile phase is described by a similar equation. The substitution in the expression for the capacity factor for monobasic bases yields the equation:

$$k = \frac{k_0 + k_1 \dfrac{[H^+]_m}{K_{a_m}}}{1 + \dfrac{[H^+]_m}{K_{a_m}}} \qquad \textbf{Equation (9)}$$

where k_0 and k_1 are respectively the capacity factors of the neutral and the fully ionized bases [46].

The Solvofobic theory has been criticized for disregarding the specific interactions between the analyte and the silanol groups of the stationary phase, the electron-donor and the elekron-acceptor interactions and the formation of hydrogen bonds. Regardless of the nature of the intermolecular interactions, in liquid chromatography dynamic equilibrium of the analyte concentration in the mobile and the stationary phase of the chromatographic column is observed and the content of the analyte in the mobile phase is a function of the pH of the medium [46].

With the progress of the science, the theory of Horvath received its confirmation and development. As mentioned above, the retention in acidic, basic and amphoteric compounds depends mainly on two of their properties: their hydrophobicity and their ability to ionize in the mobile phase. The electroneutral forms of the analytes have a relatively high affinity to the non-polar stationary phase and that is the reason why their retention is greater than the retention of the dissociated forms of the same analyte, which are more soluble in the mobile phase. The dissociation of the molecules of the analyte, which depends on the pH of the medium and on their pKa value, strongly influences the retention time of the ionized compounds [55]. As a total effect of the various intermolecular interactions, a dynamic equilibrium between the concentration of the analyte in the mobile and in the stationary phase of the chromatographic column is established. When estimating these equilibrium equations, which are mathematical expressions of the thermodynamic description of the dependence of the retention of the analytes from pH of the medium [55] were derived the following equations.

24

$$K = \frac{K_{[A-]} + K_{[HA]}(10^{pH-pKa})}{1 + 10^{pH-pKa}} \qquad \text{Equation (10)}$$

$$K = \frac{K_{[BH+]} + K_{[B]}(10^{pKa-pH})}{1 + 10^{pKa-pH}} \qquad \text{Equation (11)}$$

where $K_{[A-]}$ and $K_{[HA]}$ are the retention factors of the ionized and nonionized forms respectively for the acidic compounds, $K_{[BH+]}$ and $K_{[B]}$ are the retention factors of the ionized and nonionized forms respectively for the basic compounds [55].

The established relations between the obtained chromatographic value logK of the investigated compounds with their lipophilicity parameter logP determined by the classical method, and the differences existing between the three main approximations in determining the correlation logK/logP, supposed the development of additional methods for comparison of the results. Not taking into account that certain interactions in the theoretical models lead to obtaining accurate results only for certain groups of compounds and to deviations from the actual values in others.

Chapter II

Experimental

HPLC method for the determination of the lipophilicity paramenter and the pKa values of methylparaben, ethylparaben, propylparaben and butylparaben

Materials and methods

Reagents

Acetonitrile gradient grade, LiChrosolv-Merck, n-octanol, potassium phosphate, phosphoric acid, methylparaben, ethylparaben, propylparaben, and butylparaben are purchased from Merck and used as received.

25

Buffer Preparation

0.02 M potassium phosphate was dissolved with bidistilled water

Chromatographic Conditions

HPLC Integrated System Shimadzu LC2010 A equipped with chromatographic software Class VP 6.0 was used. RP column C18 (10 μm, 250 mm x 4.6mm i.d).

Mobile phase preparation:

Phosphate buffer (0.02 M potassium phosphate) and acetonitrile in ratio 65/35 (v/v) respectively were mixed. The different pH ranges of the mobile phase (2.0–9.45) were obtained by adjustment with phosphoric acid. As a non-retained compound, a peak of methanol was used.

Flow rate - 1.5 mL min^{-1}

Injection volume - 10 μL

Wavelength λ=254 nm.

Spectrophotometric method for determination of pKa of parabens

Experimental conditions:

The dissociation constants for the esters of 4-hydroxybenzoic acid were determined in aqueous solution with ionic strength 0.1 mol/l, which was achieved by the addition of NaCl to the suitable buffer solution. For the preparation of the solutions in the area of 4.8 to 6.0 pH units were used succinate buffers, for pH values of 6.5 to 8.2 was used a phosphate buffer, for a pH value of 8.5 to 9.7 - borate buffer, a pH range of 10.0 to 10.6 - carbonate buffers [66], and for pH 11.0 and above - upward titrated solution of NaOH. In all of the buffer solutions the concentration of sodium ions (Na^{+}) was equal 0.1 mol/l.

The pH values of the buffer solutions was determined by a pH meter METLER TOLEDO with temperature compensation. Calibration of the instrument was made with standard buffer solutions of pH 4.00, 7.00 and 9.00 produced by Merck.

The ultraviolet absorption spectra of investigated parabens were measured by using a spectrophotometer Hitachi U-3210, equipped with quartz cuvettes with a thickness of the absorbing layer 1 cm.

The spectrophotometric determination of the pK value for a given compound must contain the following steps:

1.   Investigating the spectras of the acidic (protonated) and the basic (deprotonated) forms.

For this purpose it is necessary that a spectrum of the investigated compound in 0.01 and 0.1 M solutions of hydrochloric acid and 0.01 and 0.1 M sodium hydroxide solution should be obtained. The pH of the solution mentioned above is approximately 1-2 for the hydrochloric acid, and about 12-13 for the sodium basic solutions. If the pKa value of the investigated compound is placed in the range 4-10, the recorded spectra at pH = 1-2 will be identical and will refer only to the fully protonated form of HA. The spectra at pH = 12-13 will also be equal and will refer to the deprotonated form A. If the absorption of the tested compound in both the acidic and the basic solutions for the two concentrations (0.01 and 0.1 M) differs by more than 1%, this means that the pKa value of this substance is located outside the range of 4-10 and that at a pH = 1-2 (or 12-13) both forms of the component are still present in the solution.

2.   Selection of analytical wavelength for determination of the pKa values of the spectra of acid-basic forms.

As analytical wavelength of the spectrophotometric determination was used the wavelength (λ_{anal}) where the molar absorption coefficients of the acidic and basic forms differs most from each other. When there is more than one such wavelength which is equivalent to each other the one that is closest to the maximum λ_{max} of one of the forms is chosen. All following measurements can be made at a selected wavelength λ_{anal}. The values of the absorbance of the acid-basic forms of the compound at the analytical wavelength λ_{anal} (A_{HA} and A_A) can be used in order to determine the pKa value of the tested components as described below.

3. Determination of the approximate value of pK_a.

To determine the approximate value of pKa were prepared solutions of the investigated compound in a buffer with pH values where both the forms (acidic-protonated HA and basic-deprotonated A) were presented in comparable concentrations. After determination of the absorbance of the solution at the analytical wavelength λ_{anal}, the pK value was calculated by the following equation:

$$pK_a = pH + \lg\frac{A_A - A}{A - A_{HA}} \qquad \textbf{Equation (12)}$$

where A_{HA} and A_A are the values of the absorption coefficients of the acidic and the basic forms measured at the analytical wavelength λ_{anal}. A is the absorption coefficient of the compound at a determinate pH value and **pH** is pH of the buffer solution.

4. Accurate determination of the pKa value.

After finding an approximate value of pKa 6-8 buffer solutions at pH values distributed in the range of not less $pK_a\pm0.6$, but not greater $pK_a\pm1$ were made. The determination of the absorbance of these solutions at λ_{anal} allowed the calculation of the pKa value of the investigated compound on the equation 12, shown above [66].

Stock solutions of test substances were prepared in accurate concentrations, the initial dissolving in ethanol and the subsequent dilutions in water. The working concentrations of the esters of 4-hydroxybenzoic acid were in the range of 4×10^{-5} mol/l to 8×10^{-5} mol/liter and were prepared with distilled water free from carbon dioxide.

Determination of the antioxidant activity of the investigated preservatives

Determination of the antioxidant activity of the preservatives was carried out by differential pulse voltammetry. In the proposed voltammetric approach the electroreduction of the oxygen (ER O_2) process at the mercury film electrode (MFE) is treated as a "model" reaction because of the similar processes of ER O_2 and the oxygen reduction in the tissues. It proceeds at the cathode in several stages with formation of the ROS, such as $O_2^{\bullet-}$ and $HO_2^{\bullet}$:

$$O_2 + e^- \xrightarrow{k_0} O_2^{\bullet-}$$

equation (13)

$$O_2^{\bullet-} + H^+ \rightleftharpoons HO_2^{\bullet}$$

equation (14)

$$HO_2^{\bullet} + H^+ + e^- \rightleftharpoons H_2O_2$$

equation (15)

The formation of $O_2^{\bullet-}$ is limited and controls the kinetic of the whole process. The oxygen radicals are highly reactive and toxic. They are involved in the pathogenesis of many diseases. An antioxidant reacting with $O_2^{\bullet-}$ and $HO_2^{\bullet}$ decreases their concentration at the electrode under otherwise equal conditions. The current of electroreduction of O_2 also decreases. We have suggested the following mechanism of this interaction:

$$O_2^{\bullet-} + R-C-OH \xrightleftharpoons{k_1^*} HO_2^{\bullet} + R-C=O$$

equation (16)

$$HO_2^{\bullet} + R-C-OH \rightleftharpoons H_2O_2 + R-C=O$$

equation (17)

where **R-OH** is the reduced form of the antioxidant and **R=O** is the oxidized form of the one.

The obtained results for the antioxidant activity of the investigated preservatives were compared to the results for the antioxidant activity of vitamin C, determined by the same method. The instrument utilized for this experiment was a voltammetric analyzer model TA –2 (Tomsk production). In this work the instrument in connection with PC was used. The electrochemical cell with three-electrode configuration was connected to the analyzer. A working mercury film electrode, a silver-silver chloride electrodes with KCl saturated (Ag|AgCl|KCl$_{sat}$), as reference counter electrodes, were used.

Determination of the lipophilicity parameters and pKa value of the parabens

The ability of a preservative to interact with the membrane of the microorganisms depends on its lipophilicity. This is the reason why the determination of the lipophilicity parameter is very important and is the main task in this work. In literature, log P is adopted in order to evaluate this parameter. The value of log P (approximated in the overview with log k') is successfully used for the assessment of the lipophilicity of a substance when its molecules are in the neutral form (below is described the relationship between the values of log P and log k 'of the test compounds, obtained respectively, by spectrophotometric and HPLC methos). The correct determination of k' is especially important for the correct use of the preservative.

The correct determination of the dead time of the chromatographic column (t_0) is essential for the evaluation of k', and hence for the determination of the lipophilicity parameter. The established method for determining t_0 is to pass a nonsorbed component through the chromatographic column. In this case, the dead time was recorded as the retention time of ionic compound (KBr) in nonpolar stationary phase of the chromatographic column C_{18}.

It is possible, however, depending on the environmental conditions, for the molecules of the test compound to obtain a full or a partial charge (to be protonated or deprotonated). In this case the evaluation of lipophilicity of the analyte used in literature is log D. D is also known as "apparent partition coefficient" [49].

Parameter log D describes the pH dependence of the distribution of the preservative between the membrane and the surrounding environment. Log D depends on the pKa value of the preservative. In some cases when there is a change in the polarity of the molecules of the preservative log D is the parameter for assessing the ability of a

30

compound to penetrate the cell membrane of the microorganism. The presence of certain functional groups in the molecules of the preservative provokes the processes of partial or complete protonation or deprotonation and changes its lipophilicity. In determining the pKa of parabens were used the difference in the retention times (respectively log k') between the neutral and ionized forms of investigated compounds. This transition in the molecules structure from the neutral to the ionized state is carried out in the pH range close to the pKa value of the test compound, so when we determine the value of the pH at the range where the rate of transition is highest, we find the point from the curve which corresponds to the pKa value of the compound.

The graphs in Figure 4 represent the transition from log P and log D and show how the dependence of the lipophilicity on the pH of the medium changes. For the components with one ionizing group the graph "partition coefficient/pH" is usually a sigmoid curve [49]. The inflection point of the curve corresponds to pKa value of the investigated compound [49]. The graph on the dependence of log k'/pH for the parabens (substances having one ionizable group) is very similar to that described in literature as the "partition coefficient/pH" for this type of compounds. All approximations which describe well enough this dependency are sigmoidal curves. Among the best fits, the equation of asymmetric sigmoid was selected as the most suitable. The reason to choose this equation was that it adequately enough (R^2=0.996) describes the obtained relation, and on the other hand, compared with other derived equations, it is relatively simple. (Figure 4).

From the data in Figure 4 the mathematical dependency on (log k ') from pH for methylparaben, ethylparaben, propylparaben and butylparaben was determined. This dependency is described by equation 18, and its coefficients are shown in Table 2.

$$y = C + \frac{A}{[1 + \exp((x - D) * K)]^n}$$

$$K \neq 0; 20 \leq n \leq 100$$

Equation (18)

Figure 4. Curves of pH dependency of the lipophilicity parameter log k' of methylparaben (MP), ethylparaben (EP), propylparaben (PP), butylparaben (BP).

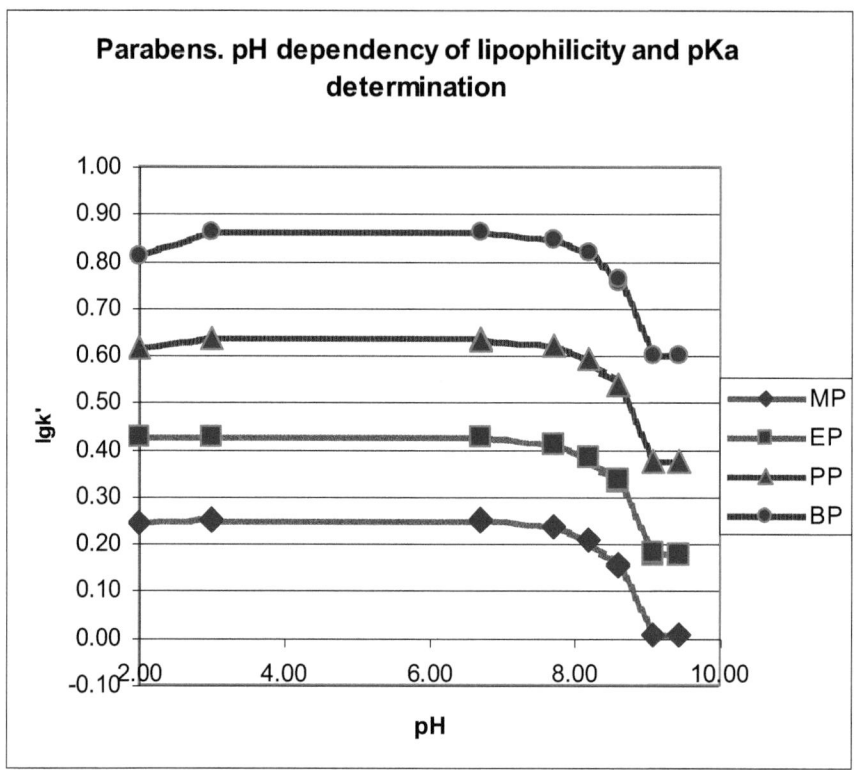

Table 2. Correlation coefficients of the equation described the dependency of (log k') from the pH of the media for parabens and calculated with equation 18 pKa values.

	A	K	C	D	R^2	Obtained pKa
MP	0.2447	3.5464	0.0009	10.0923	0.996	8.87
EP	0.2518	3.6068	0.1683	10.0848	0.995	8.90
PP	0.2598	3.7263	0.3686	10.0347	0.996	8.87
BP	0.2643	3.6452	0.5913	10.0588	0.997	8.79

There is no data for the lipophilicity parameter of parabens obtained by HPLC method published in the specific literature. The pKa values for the investigated compounds are not determined with chromatographic methods.

Calculating the second derivative of the equation of the dependence of (log k') of the pH of parabens can determine the inflection point, which in this case coincides with the pKa value of the investigated compounds. The results for the pKa of this experiment are compared in Table 3 with the literature data for these compounds determined with the spectrophotometric method for determination of pKa, and pKa data for the same compounds were obtained with equation 20, which was derived from well-known in the literature equation 19:

$$k = \frac{k_{HA} + k_A * 10^{pH-pKa}}{1 + 10^{pH-pKa}} \qquad \text{Equation (19)}$$

$$pKa = pH - \lg\frac{k_{HA} - k}{k - k_A} \qquad \text{Equation (20)}$$

where k is the retention factor for the investigated compound at a determined level of pH, k_{HA} and k_A are the retention factors of the neutral and the fully ionized form of the investigated compound, pKa is the pKa value of the investigated compound and pH is the pH of the medium [62].

The pKa values in Table 3, obtained by HPLC method are close to the literature data, which shows that the chromatographic conditions are well chosen and gives us the opportunity to obtain the correct results for the pKa values of the investigated parabens. The data for the pKa values of the substance can be obtained by substituting the corresponding values of k' in the equations 19 and 20. The results for the pKa values were obtained from the second derivative of Equation 18 and are closer to the literature data than those obtained by the equation 20. It can be concluded that the use of the most common stationary phase C18 is suitable for determining the coefficient of lipophilicity of esters of paraben. Another advantage of the newly developed liquid chromatographic method is that it enables the analyst to choose the most appropriate mathematical approximation for the test compound.

Table 3. Comparison between the data from the experiment and the literature data for the pKa value of the parabens.

	pKa value obtained with equation 21	pKa value obtained with equation 19	pKa value obtained with spectrophotometric method (equation 12)	pKa values in the literature
MP	8.87±0.02	8.95±0.02	8.40±0.02	8.47
EP	8.90±0.02	9.16±0.02	8.38±0.02	8.50
PP	8.87±00.2	8.87±0.02	8.31±0.03	8.47
BP	8.79±0.02	8.94±0.02	8.38±0.06	8.47

Two consequitve and opposite in nature processes are observed in the molecules of parabens. The first one, which is significantly weaker, is the partial protonation of the oxygen atom of the ester group (seen in the low-pH values in the first part of the graphs Figure 4) particularly in long-chain homologues. The second process is the deprotonation of the hydroxyl group (reflected in the second part of the graphs in Figure 4).

The process of protonation is very weak and highly dependent on the length of the aliphatic group in the side chain of the ester group. That can also be registered at the curves log k'/pH, especially for the long chain homologues butyl and propyl-parabens at pH ranges lower than 3 (Figure 5). It is well known that the longer the aliphatic chain of the ester is, the stronger its positive induction effect is (electron donors properties increases). That increases the electron density at the oxygen of the $C = O$ group and therefore its ability to coordinate protons, especially at lower pH ranges, increases.

The process of deprotonation of the hydroxyl group is much stronger and more important for the characteristics of the preservatives (lipophilicity parameters and pKa values). The inflection points in the curves, which correspond to the determined

34

pKa values of the parabens with this method, are very close to their literature pKa values, and are in excellent agreement with the values obtained with the wellknown equation 20.

Determining the dependence lg k'/pH of parabens in pH values of the mobile phase above 8 for the duration of the experiment has not registered changes in the behavior of the chromatographic system due to spoiling of the column. In literature there are described many cases where the chromatographic column of the same type as described above can be reliably used at pH 9 of the mobile phase for about 500 injections [67] without significantly affecting the quality of the analysis. After that, it is not recommended that this column should be used for the quantitative analysis as a front in the peaks of the analyzed compounds appears [66]. However, this does not affect the retention times of the analyts and the column can be used for qualitative analyzes and determinations.

Figure 5. Protonaton and deprotonation of the parabens molecules at different pH ranges.

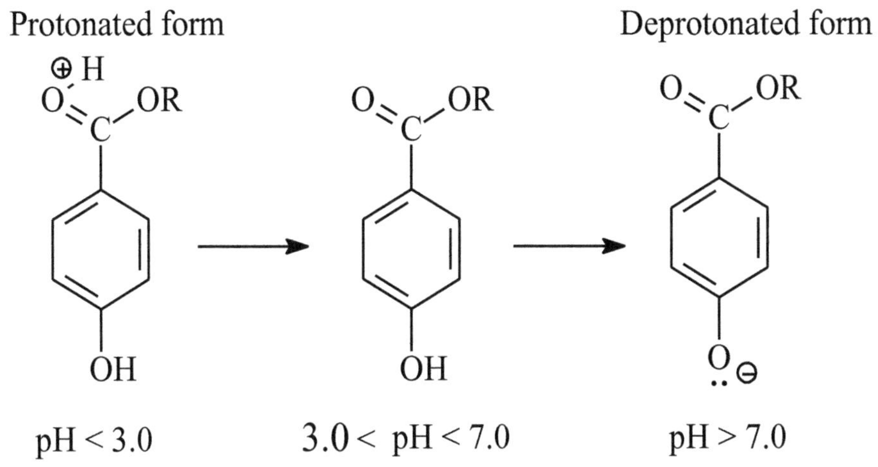

Protonated form   Deprotonated form

pH < 3.0   3.0 < pH < 7.0   pH > 7.0

To evaluate the correctness of the HPLC method, the obtained results for the pKa and log P values of parabens were compared with the results obtained with the spectrophotometric method alternative to chromatographic determination. As distinguished from the chromatographic method, the spectrophotometric determination of the pKa of a given compound is possible, if the ionisation is not complicated by the side (additional) processes and the following conditions are accomplished [66]:

• The coefficient of molar absorption for the both forms, the acidic and the basic (protonated and deprotonated) do not depend on the pH of the solution.

• the value of the pKa of the test compound is in the pH range between 2-11.

• The absorptions in the solutions containing only one of the forms of acid-base balance (only protonated A_{HA} or deprotonated A_A forms) can be experimentally determined.

• In the spectra of the acidic and basic forms exist an area range in which their coefficients of molar absorption vary markedly.

The absorbtion spectra of propylparaben are shown in Figure 6.

Figure 6: Absorbtion spectra of propylparaben (5.0.10-5 M) in water.

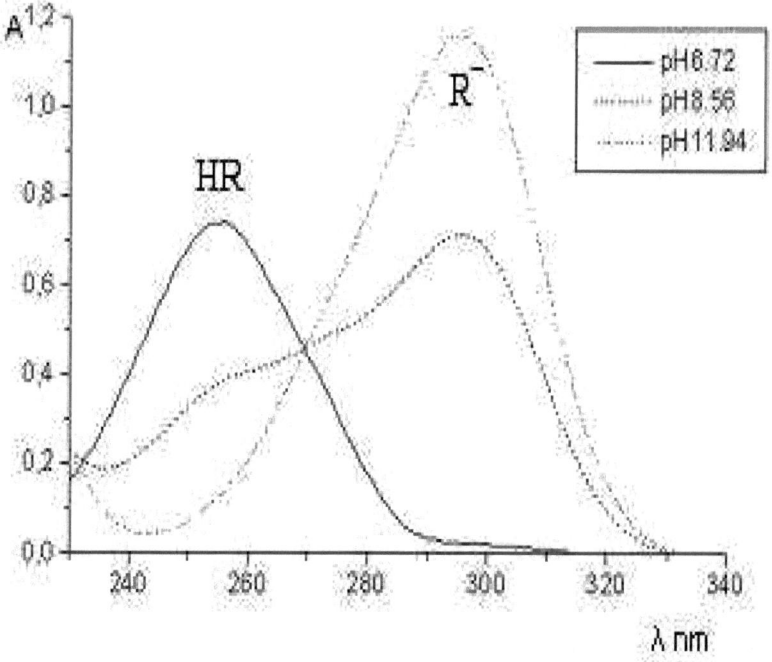

As could be seen from the procedure described above, the HPLC method for the determination of pKa value is distinguished by its accessibility and effectiveness (ability to determine the pKa value of more than one component simultaneously). Comparing the chromatographic equation (20) and spectrophotometric equation (12) for calculating the pKa, they are almost identical as a mathematical record, but use different physico-chemical parameters of the investigated compounds which are affected by the pH of the medium in an identical way. These parameters are k' (retention factor) in liquid chromatography and A (absorption coefficient) in spectrophotometry. The results obtained by the spectrophotometric method are more accurate and precise than those obtained by the chromatographic method, but the spectrophotometric method is very slow and labor intensive. Other achievement of the chromatographic method is that it allows simultaneous determination of the lipophilicity parameters of all parabens.

The distribution coefficient of parabens log P was calculated from data obtained by the classical method for determining the lipophilicity parameter with spectrophotometric determination of the distribution of the components in the system n-octanol/water. The concentration of the different components was determined at a wavelength of 254 nm. Log P values were calculated using equation (21). The conditions for the spectrophotometric determination was as close as possible to the conditions for liquid chromatographic determination (pH 3.0 buffer solution to the system octanol/buffer concentration of buffer 0.02 M potassium phosphate, an extraction temperature of 37 ° C, $\lambda = 254$ nm).

$$\lg P_{O/W} = \lg C_O - \lg C_W \quad \textbf{Equation (21)}$$

where C_o and C_w are respectively the concentrations of the parabens after the distribution in the n-octanol/water system in the octanol part and in the water part. The Log P data are presented in Table 4.

The relation found between the value of log P (Table 4) obtained by liquid-liquid extraction system in the n-octanol/buffer, and the value of the log k ', obtained by the

HPLC method under the conditions described above, has got a very good correlation (R2 = 0.9981 Figure 7).

Figure 7. Dependency log k' on log P

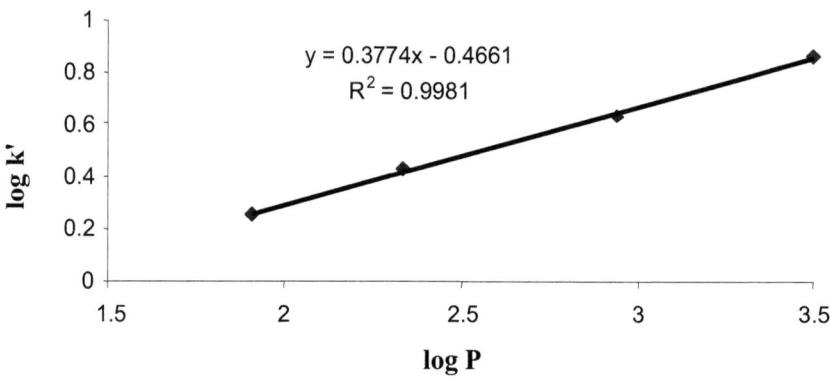

Table 4. Log P values for parabens at T°C=37 and pH 3 obtained with liquid/liquid extraction.

Methylparaben	Ethylparaben	Propilparaben	Butylparaben
1.91±0.01	2.343±0.001	2.94±0.01	3.50±0.02

The maintaining of the most suitable pH range, in which preservatives are most active, should be taken into account. It is well known that in the pH value above 8.0 the process of alkaline hydrolysis is induced in the molecules of the parabens. This yields the corresponding alcohol and p-hydroxybenzoic acid. Even at pH about 7.0, the hydrolysis processes are already noticeable [17][12]. This is the reason why the use of parabens in products in the pH range above 7.0 should be avoided.

Through defining the lipophilicity parameters of the parabens with the methods of the high performance liquid chromatography and investigating the correlation between the log k' and the pH, the pH range of the best effectiveness of the parabens as preservatives was determined.

Figure 8: Dependency of the lipophilicity parameters of log k' of the number of carbon atoms in the side chain at the parabens.

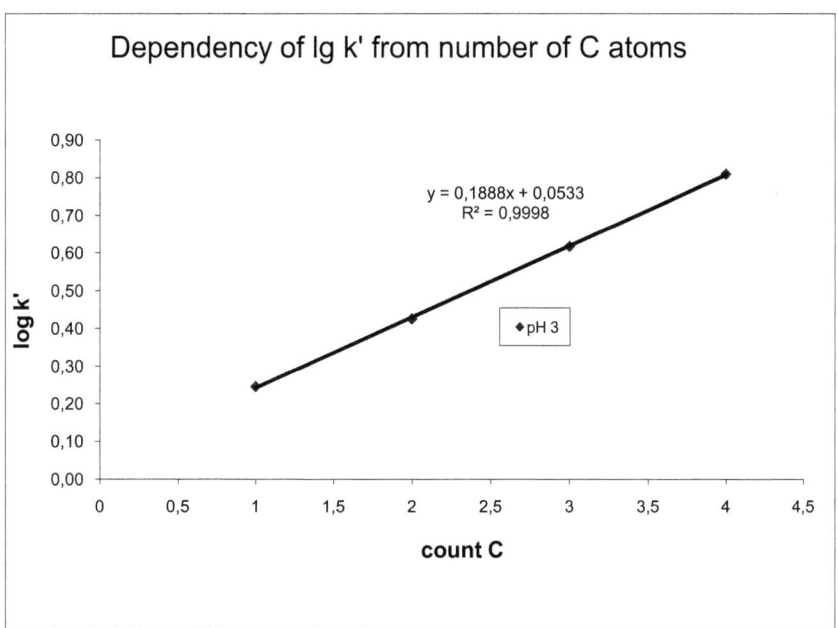

The observed correlation between pH and the log k' clearly indicates that there is an optimal pH range for the use of parabens as preservatives. For the esters of p-hydroxybenzoic acid this range is at pH values lower than their own pKa. In this range their lipophilicity parameter is the largest and provides the best penetration of the preservatives molecules through the bacterial cell membrane. According to the obtained results the optimum range for parabens pH is from 3 to 6.5.

When increasing the number of carbon atoms in the side chain of the parabens their lipophilicity parameter increases linearly (Figure 8).

Determination of the total antioxidant activity

In order to investigate the antioxidant activity, the voltammograms of electoreduction of oxygen (O_2) current were recorded as a function of potential at the

39

working electrode (mercury film electrode) in the supporting electrolyte containing the investigated substances.

Total antioxidant activity coefficients of the preservatives Table 5.

Preservative	Concentration in g/l = 0.05	Concentration in g/l =0.5	Concentration in g/l = 5
Coefficient of total antioxidant activity (K μmol l^{-1} min^{-1})			
Methylparaben	0.76	0.79	1.01
Ethylparaben	0.79	0.84	1.08
Propylparaben	0.83	0.95	1.31
Butylparaben	0.87	0.97	1.43
Ascorbic acid	0.95	1.30	1.42

All of the investigated preservatives show antioxidant activity comparable with the antioxidant activity of ascorbic acid. The compound with the highest antioxidant activity is butylparaben, and with the lowest – methylparaben, or the antioxidant activity increases with the increase of the substituent in the ester group (methyl <ethyl <propyl <butyl). The lipophilicity parameters of parabens increases in the same sequence.

The comparison of the total antioxidant activity with lipophilicity parameters of parabens is presented in **Table 6**.

Table 6: Comparison of the coefficient of total antioxidant activity K with the lipophilicity parameter log k' of parabens.

Preservative	Coefficient of total antioxidant activity (K μmol l^{-1} min^{-1})	Lipophilicity parameter log k'	Lipophilicity parameter log P
Methylparaben	0.76	0.25	1.91
Ethylparaben	0.79	0.43	2.34
Propylparaben	0.83	0.62	2.94
Butylparaben	0.87	0.83	3.50

As we can clearly see from the Table 6, the coefficients (K, log k' and log P) have increased with the increasing of the molecular weight of the parabens. Generally, in this case it can be concluded that the substances with a high degree of lipophilicity are the better antioxidants.

Figure 9. Dependency of the coefficient of total antioxidant activity (K μmol l⁻¹ min⁻¹) on log k'

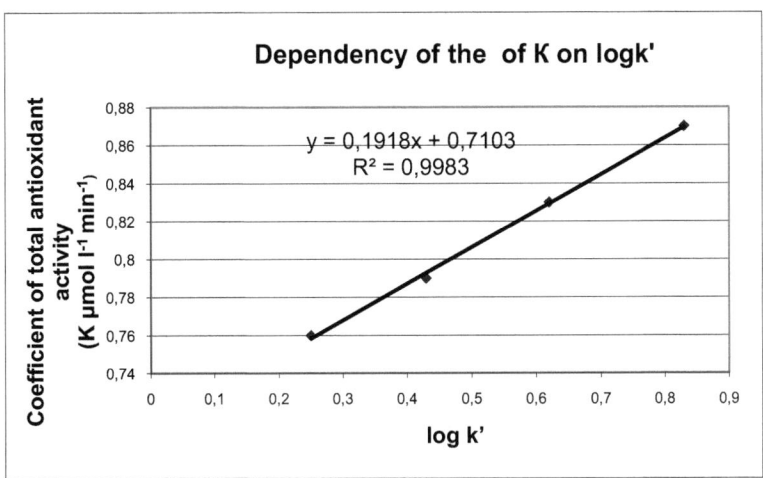

Figure 10. Dependency of the coefficient of total antioxidant activity (K μmol l⁻¹ min⁻¹) on log P

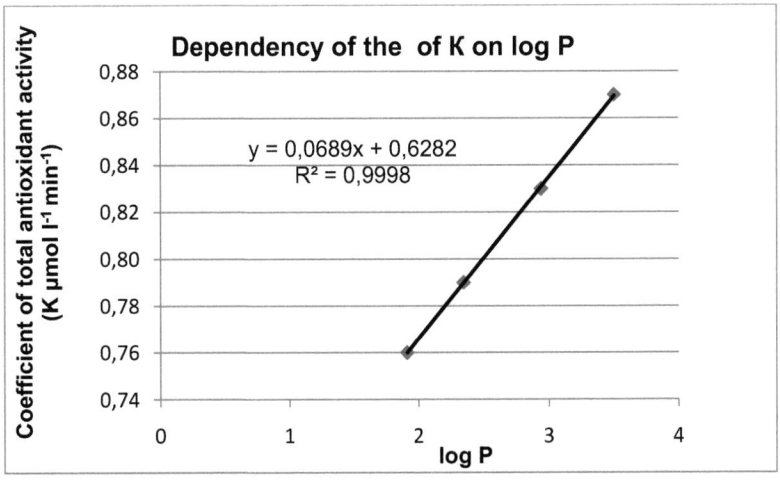

The practical calculations of lipophilicity of the parabens, as log P and log k`, and determination of the total antioxidant activity coefficients as K, which were performed indicated that a relationship between their lipophilicity and antioxidant activity exists. This dependency is linear and with excellent correlation, respectively 0.9998 for the relationship K/logP (Figure 10) and 0.9983 for the relationship K/log k'(Figure 9). At almost double increasing of the lipophilicity of the parabens the antioxidant activity increases from methyl to butyl paraben with about 15%.

References:

[1] Donald Jzconia, Preservatives in pharmaceutical products, In "Quality control in the parmaceutical industry" edited by Murray S. Cooper, Academic press New York San Francisco London 1972 volume I pp.102-126

[2] H. Lord, J. Pawliszyn. Journal of Chromatography A, 902, (2000), 17–63.

[3] Pölönen, I. 2000 In: Animal Science Preservation Efficiency of Formic Acid and Benzoic Acid in the Ensiling of Slaughterhouse By-Products and their Subsequent Metabolism in Farmed Fur Animals. pp. 7-8 Ph.D. dissertation, University of Helzinki Department of Animal Science Helsinki

[4] H. Yamazaki, T. Yoneda, T. Yamaguchi. Jpn. J. Food Chem, Vol. 5(2), 1998

[5] H. Lin, Y.Choong. Journal of Food and Drug Analysis, 7(4), (1999), 291-304

[6] Legislation, intake and usage of food additives in Ireland

http://www.fsai.ie/publications/reports/Legislation_Food_Additives.pdf

[7] M. Johnston, G. Hanlon, S. Denyer, R. Lambert. Journal of Applied Microbiology, 94, (2003), 1015–1023

[8] R. Francisco, B. Fernando, J. Frades, P. Buxeda. Wat. Res. 35, (2), (2001), 378-396

[9] J. Beltran-Heredia, J. Torregrosa, J. Dominguez, J. Pers. Chemosphere, 42, (2001), 351-359

[10] L. Ste-Marie, D. Boismenu, L. Vachon, J. Montgomery. Analytical Biochemistry 241, (1996), 67-74

[11] G. Casini, N. De Laurentis, N. Maggi, S. Ottolino, Farm. Ed. Prat., **36,** (1981), 553-558.

[12] M. Soni, S. Taylor, N. Greenberg, G. Burdock, Food and Chemical Toxicology, 40, (2002), 1335-1373

[13] Y. Nakagawa, G. Moore. Biochemical Pharmacology, 58, (1999), 811-816

[14] N. Valkova, F. Lepine, L. Valeanu, M. Dupont, L. Labrie, J. Bisaillon, R. Beaudet, F. Shareck, R. Vilemur. Applied an Environmental Microbiology, June, 67(6), (2001), 2404-2409

[15] S. Brul, P. Coote. International Journal of Food Microbiology 50, (1999), 1-17

[16] P. Furrer, J. Mayer, R. Gurny. European Journal of Pharmaceutical and Biopharmaceutical 53, (2002), 363-280

[17] M. Soni, G. Burdock, S. Taylor, N. Greenberg. Food and Chemical toxicology 39, (2001), 513-532

[18] The Flavor And Fragrance High Production Volume Consortia
The Aromatic Consortium Test Plan for Benzyl Derivatives Submitted to the EPA under the HPV Challenge Program by: The Flavor and Fragrance High Production Volume Chemical Consortia- AR 201-13450A

[19] E. Charlesworth, Annal Sofallergy, Asthma, & Immunology, 76, JUNE, (1996), 484-496.

[20] G. Michaelsson, L. Juhlin. Br.J. Dermatol, 88, (1973), 525–32

[21] E. Macy, M. Schatz, R. Zeiger. The Permanente Journal/Fall 2002/ Volume 6, No 4, 17-21.

[22] J. von Elbe, J. Schwartz, R. Lindsay (1996) In: Food Chemistry. ed. Marcel Dekker INC, New York - Basel - Hong Kong, pp. 782-792

[23] Comprehensive Reviews in Food science and Food Safety-Chapter III Factors that Influence Microbial Growth Vol. 2 (Supplement), 2003

[24] O. Padilla Zakour, Chemical Food Preservatives: Bonzoate & Sorbate, Venture A Quarterly Newsletter Published by New York State Food Venture Center Geneva, NY   Summer   1998·Vol.   1   No.   2 (http://www.nysaes.cornell.edu/necfe/pubs/pdf/Venture/venture2_chemical.html)

[25] N.Veien, T. Hattel, G. Laurberg. Contact Dermatitis 34 (6), (1996), 433

[26] K. Bajaj, A. Chatterjee. Indian Journal of Dermatology Venereology and Leprology, 51 (6), (1985), 319–321

[27] S. Oishi. Food and Chemical Toxicology, 40, (2002), 1807-1813

[28] R. Di Giovannandrea, L. Diana, M. Fiori, E. Ferretti, G. Foglietta, R. Caronna, G. Severinia. Journal of Chromatography B, 751, (2001), 365–369

[29] R. Driouich, T. Takayanagi, M. Oshima, S. Motomizu. Journal of Chromatography A, 903, (2000), 271–278

[30] J. Koundourellis, E. Malliou, T. Broussali. Journal of Pharmaceutical and Biomedical Analysis, 23, (2000), 469–475

[31] C. Garcia, A. Breier, M. Steppe, E. Schapoval, T. Oppe. Journal of Pharmaceutical and Biomedical Analysis, 31, (2003), 597-600

[32] D. Kreuz, A. Howard, D. Ip. Journal of Pharmaceutical and Biomedical Analysis, 19, (1999), 725-735

[33] J. Rauha, H. Salomies, M. Aalto. Journal of Pharmaceutical and Biomedical Analysis, 15, (1996), 278-293

[34] D. Kollmorgen, B. Kraud. Journal of Chromatography B, 707, (1998), 181–187

[35] S. Kang, H. Kim. Journal of Pharmaceutical and Biomedical Analysis, 15, (1997), 1359-1364

[36] M. Akhtar, S. Khan, I. Roy, I. Jafri. Journal of Pharmaceutical and Biomedical Analysis, 14, (1996), 1609-1613

[37] G. Shabir. Journal of Pharmaceutical and Biomedical Analysis, 34, (2004), 207–213

[38] M. Tomassin, E. Cavalli, Y. Guillaume, C. Guinchard. Journal of Pharmaceutical and Biomedical Analysis, 15, (1997), 831-838

[39] M. Collado, V. Mantovani, H. Goicoechea, A. Olivieri. Talanta, 52, (2000), 909-920

[40] C.Palmer. Journal of Chemical Education, 76 (11), (1999), 1542-1543

[41] M. Blanco, J. Coello, H. Iturriaga, S. Maspoch, M. A. Romero, Journal of Chromatography B, 752, (2001), 99–105

[42] M. Blanco, J. Coello, H. Iturriaga, S. Maspoch, M. Romero. Journal of Chromatography B, 751, (2001), 29–36

[43] L. Labat, E. Kummer, P. Dallet, J. P. Dubost. Journal of Pharmaceutical and Biomedical Analysis, 23, (2000), 763–769

[44] Q. Xu, L. Trissel. "Stability-Indicating HPLC Methods for Drug Analysis", published by the American Pharmaceutical Association (1999), 111-243

[45] Хр. Димитров, Н. Пецев. Газова Хроматография, София 1974, Наука и изкуство. Стр. 383

[46] C. Horvath, W. Melander, I. Molnar. Analytical Chemistry, 49, No1, January 1977

[47] P. Wiczling, M. Markuszewski, R. Kaliszan. Anal. Chem. 76, 2004, 3069-3077

[48] T. Braumann. J. Cromatogr. 373, (1986), 191-225

[49] H. Van de Waterbeemd, B. Testa, Advanced in Drug Research, 16, (1987), 85-225

[50] C. Hansch, T. Fujita. Journal of the American chemical society, 86, (1964), 1616

[51] S. Bhal. Application Note, Advanced Chemistry Development, Inc. Toronto, ON, Canada, (http://www.acdlabs.com/download/app/physchem/logp_vs_logd.pdf)

[52] S. Bhal, Application Note. Advanced Chemistry Development, Inc. Toronto, ON, Canada, (http://www.acdlabs.com/download/app/physchem/making_sense.pdf)

[53] A. Leo, C. Hansch, D. Elkins. Chemical Reviews, 71, (6), 1971

[54] M. Ramos-Nino, M. Clifford, M. Adams. J. Appl. Bacteriol., 80, (1996), 303–310

[55] R. Kaliszan, P. Wiczling. Anal Bioanal Chem, 382, (2005), 718-727

[56] R. Kaliszan. Journal of Chromatographic science, 22, (1984), 362-370

[57] K. Valko, P. Slegel, J.Chromatogr., 631, (1993), 49-61

[58] T. Slawik, C. Kowalski, J Chromatogr. A, 952, (2002), 295-299

[59] R. Kaliszan, P. Haber, D. Siluk, K. Valko. J Chromatogr. A, 965, (2002), 117-127

[60] R. Kaliszan, P. Wiczling, M. Markuszewski, Anal. Chem., 76, (2004), 749-760

[61] P. Wiczling, M. Markuszewski, R. Kaliszan. Analytical Chemistry, 77(2), (2005), 449-458

[62] I. Canals, J. Portal, E. Bosch, M. Roses. Anal Chem., 72, (2000), 1802-1809

[63] A. Avdeef, J. Comer, S. Thomson. Anal. Chem., 65, (2003) 42-49

[64] A. Avdeef, Current Topics in Medicinal Chemistry, 1, (2001), 277-351

[65] D.D. Perrin, Aust. J. Chem. (1963), 16, 572-578.

[66] L. Loginova, O. Chernisheva, A. Vlasenko, A. Kulikov, Ya. Atamanichenko, Kharkov University Bulletin. 13, (2005), 36

[67] J. Dolan, LC GC Europe, 21 (11), (2008), 562-567